FORSCHUNGSBERICHTE DES LANDES NORDRHEIN-WESTFALEN

Nr. 2282

Herausgegeben im Auftrage des Ministerpräsidenten Heinz Kühn
vom Minister für Wissenschaft und Forschung Johannes Rau

Dr. Michio Matsumoto *

Institut für Theoretische Physik der Ruhr-Universität Bochum

Cross-Magnetic Turbulent Diffusion of Charged Particles in a Weakly Ionized Plasma*)

Westdeutscher Verlag Opladen 1972

*) Work performed within the activities of the Sonderforschungsbereich "Plasmaphysik Bochum/Jülich"

* On leave from Tokyo Institute of Technology

ISBN 978-3-531-02282-6 ISBN 978-3-322-87851-9 (eBook)
DOI 10.1007/978-3-322-87851-9

Gesamtherstellung: Westdeutscher Verlag

Contents

Introduction

It is very important to know how field fluctuations affect the diffusion of plasma across a strong magnetic field $\underline{B}$. Many investigators have studied this problem theoretically[1)~8)] and experimentally.[9)~12)] In cases of weak turbulence, the ion diffusion coefficient $D^i_\perp$ is proportional to the electric field fluctuation energy or the mean square amplitude of density fluctuations.[1)] For a higher degree of fluctuation amplitudes exceeding a certain threshold which is given by the relation $<|\delta E|^2>^{1/2} e\, h_e/Te = \varkappa \rho_{Te}$ Dupree[7)] proposed $D^i_\perp \propto <|\delta E|^2>^{1/2}$. Here, $e = |e_e|$ is the electronic charge, h_e the electron Debye length, Te the electron temperature in unit of energy, ρ_{Te} the electron mean Larmor radius and $\varkappa = n_o^{-1} \partial n_o/\partial x$, $n_o(\underline{r})$ being the mean density of charged particles. Dupree's theory, however, is applicable only to the ion diffusion in a collisionless plasma. Walsh et al[12)] determined from measurements in the Q-machine the diffusion coefficient as a function of the mean square amplitude of density fluctuations. Their experimental results are in good agreement with the predictions of Dupree's theory.

The turbulent diffusion in the positive column was first studied by Kadomtsev.[2)] Under the assumption that Prandl's mixing length is constant, he derived the radial density distributions $n_o(\underline{r})$ in the turbulent state contracting toward the axis in comparison with the distribution in the state of classical diffusion.[3)] In contrast with his result, Matsumoto[8)] derived rather flat distributions $n_o(r)$ assuming $<\delta n \delta E_o> \propto r n_o(r)$ for the cross moments, which seems to be confirmed by experiments.[13)] The above two theoretical results are based on the quite different assumptions. The study of turbulent diffusion in a weakly ionized plasma needs more basic considerations on the correlation of density fluctuations.

Therefore, we study the turbulent diffusion in the weakly ionized plasma in a strong magnetic field $\underline{B}$ with a parallel external electric field $\underline{E}$ more rigorously. The plasma has a small gradient of the charged particle density perpendicular to the external fields. In this case, the plasma is unstable to the current convective instability.[2)] Turbulent flows of the charged particles are discussed on the basis of the correlation theory of fluctuations.[14)] The diffusion coefficients are derived in a form similar to Bohm diffusion type. The electric field fluctuations arise from the phase difference between ion and electron density fluctuations. These fluctuating fields are determined by using the Poisson equation. Then, considering only ion and electron density fluctuations, we calculate the correlations of these density fluctuations starting from the Klimontovich equations.[14)]

The outline of the calculations is described in the next section. We assume the fluctuations to be stationary and homogeneous. In Sec. 3, we derive an expression for the cross-magnetic diffusion flows using the first velocity moments of the averaged Klimontovich equations. A hierarchy for the moments of fluctuating densities is constructed in Sec. 4. In Sec. 5, we first estimate the diffusion coefficients due to small amplitude fluctuations, in which case the third and higher order moments are completely negligible. In Sec. 6, we consider the current convective instability. In Sec. 7, the diffusion coefficients in the strong turbulence case are obtained by applying an approximation to the fourth order moments in the hierarchy according to Millionshchikov. In the last section we discuss the diffusion in the intermediate region between small and strong turbulent fluctuations.

Basic Consideration

The charged particles in our plasma are exposed to many collisions by neutrals. These collision frequencies ν_a are generally of the same order or greater than the plasma frequencies π_a, where the subscript a stands for the type of the charged particles. Frequencies of the fluctuations are much less than ν_a. Hence, we assume that the charged particles have almost nearly a local Maxwellian velocity distribution. The charged particles are further affected by electric field fluctuations.

The above two kinds of effects on the charged particles are quite different, and they are considered mutually independent. We can, therefore, treat the flows of charged particles as a sum of two parts.[15)]

$$\underline{\Gamma}^a_\perp = \left\{ \underline{\Gamma}^a_\perp \right\}_{coll} + \left\{ \underline{\Gamma}^a_\perp \right\}_{fluct} \tag{1}$$

The first term $\left\{ \underline{\Gamma}^a_\perp \right\}_{coll}$ represents the cross-magnetic flows due to classical collisions by the neutrals and the second term $\left\{ \underline{\Gamma}^a_\perp \right\}_{fluct}$ those due to the fluctuations. In strong fields $\underline{B}$, the former flows decrease in proportion to B^{-2}, whereas the latter flows are expected to be proportional to B^{-1} as in the type of Bohm diffusion. Experimental investigations of the positive column in a strong magnetic field show that the flow $\left\{ \underline{\Gamma}^a_\perp \right\}_{fluct}$ is much greater than the flow $\left\{ \underline{\Gamma}^a_\perp \right\}_{coll}$.

The calculation is made in Cartesian coordinates. The external electric field $\underline{E}$ and magnetic field $\underline{B}$ are both uniform and parallel to the z-axis. The mean density gradient of the charged particles is assumed to be weak and in the x-direction only. In this case, the flow $\left\{ \underline{\Gamma}^a_\perp \right\}_{fluct}$ is derived in a form proportional to the mean density gradient. Then, we can obtain the diffusion coefficient which may be expressed as

$$D_{\perp\, fluct} = A \; \rho_{Te} \; \upsilon_{Te} \,. \tag{2}$$

Here, A is a numerical factor, $\rho_{Te} = \upsilon_{Te} / \Omega_e$ the mean Larmor radius of electrons, $\upsilon_{Te} = \sqrt{Te/m_e}$ the electron thermal velocity and $\Omega_e = e\,B/m_e\,c$ the electron Larmor frequency. Bohm proposed $A = 1/16$ without theoretical explanation. In the turbulent state, of course the values of A depend on the characteristic quantities of fluctuations. We are interested in these values.

If the microscopic state of the plasma at time t is completely determined by the well-known coordinates and velocities of particles, we can start from the Klimontovich equations for the phase densities Na $(\underline{r}, \underline{v}, t)$.

$$\frac{\partial}{\partial t} Na + \underline{v} \cdot \frac{\partial}{\partial \underline{r}} Na + \frac{e_a}{m_a} \left\{ \underline{E}^M + \frac{1}{c} \underline{v} \times \underline{B}^M \right\} \cdot \frac{\partial}{\partial \underline{v}} Na = San^{16)} \tag{3}$$

Here, $\underline{E}^M$ and $\underline{B}^M$ are the total electric and magnetic fields and San denotes the collision effects on the a-type particles by the neutrals. We neglect the collisions between charged particles. The phase density Na $(\underline{r}, \underline{v}, t)$ is a dynamical function of the coordinates $\underline{r}_{ai}$ and velocities $\underline{v}_{ai}$ of all a-type particles and is defined by

$$Na (\underline{r}, \underline{v}, t) = \sum_{1 \le i \le N_o^a} \delta(\underline{r} - \underline{r}_{ai}(t)) \; \delta(\underline{v} - \underline{v}_{ai}(t)), \tag{4}$$

where N_o^a is the total number of the a-type particles.

It is very difficult to make direct use of the microscopic equation (3). Therefore, we discuss the ensemble average over the particle positions and velocities. Averaging eq. (3) in this way, we obtain

$$\frac{\partial fa}{\partial t} + \underline{v} \cdot \frac{\partial fa}{\partial \underline{r}} + \frac{e_a}{m_a} \left\{ \underline{E} + \frac{1}{c} \underline{v} \times \underline{B} \right\} \cdot \frac{\partial fa}{\partial \underline{v}} = \overline{San} - \frac{e_a}{m_a} \frac{\partial}{\partial \underline{v}} \cdot \overline{\delta Na \delta \underline{E}}, \tag{5}$$

where fa $(\underline{r}, \underline{v}, t) = \overline{Na (\underline{r}, \underline{v}, t)}$ is the one particle distribution function given as the ensemble average of the phase density Na $(\underline{r}, \underline{v}, t)$. The fields $\underline{E}$ and $\underline{B}$ are the averaged quantities of the corresponding microscopic fields $\underline{E}^M$ and $\underline{B}^M$, and are equal to the external fields. The r. h. s. of eq. (5) denotes the collision effect for the a-type particles in a broader sence. It consists of two parts: the first is the Boltzmann type collision term $\overline{San} = (\delta fa/\delta t)_n$ and the second is a Fokker-Plank type one, where $\delta Na = Na(\underline{r}, \underline{v}, t) - fa(\underline{r}, \underline{v}, t)$ and $\delta \underline{E} = \underline{E}^M (\underline{r}, t) - \underline{E}$. If we extend the averaging procedure to the case of turbulent fluctuations, the same form of eq. (5) is held by using δfa instead of δNa, where δfa is the fluctuation part of distribution function.

In steady state the charged particle flows defined by $\underline{\Gamma}^a = \int d^3\underline{v}\, \underline{v}\, fa(\underline{r}, \underline{v})$ are calculated from equations, which we get by multiplying eq. (5) with the velocity $\underline{v}$ and by integrating over the $\underline{v}$-space. The diffusion flows, i. e. the x-components of the flows $\underline{\Gamma}^a$ are expressed as

$$\Gamma_x^a = - \frac{1}{1+S_a^2} \frac{1}{\nu_a} \frac{\partial}{\partial x} \int d^3\underline{v}\, v_x^2\, fa(\underline{r}, \underline{v}) + \frac{S_a^2}{1+S_a^2} \frac{c}{B} \overline{\delta n_a(\underline{r}, t)\, \delta E_y(\underline{r}, t)} +$$
$$- \frac{Sa}{1+S_a^2} \frac{1}{\nu_a} \frac{\partial}{\partial x} \int d^3\underline{v}\, v_x v_y\, fa(\underline{r}, \underline{v}) + \frac{Sa}{1+S_a^2} \frac{c}{B} \overline{\delta n_a(\underline{r}, t)\, \delta E_x(\underline{r}, t)} \tag{6}$$

Here, $Sa = \Omega a / \nu_a$ is the ratio of the Larmor frequency Ω_a to the collision frequency ν_a and $\delta n_a(\underline{r}, t) = \int d^3\underline{v}\, \delta Na(\underline{r}, \underline{v}, t)$ are the fluctuating densities. We have assumed that the neutral particles are uniformly distributed and the Boltzmann type collision term is approximately given by

$$\bar{S}an = \frac{\partial}{\partial \underline{v}} \left\{ v_a \underline{v} \; fa\,(\underline{r}, \underline{v}) \right\} \quad ^{16)} \tag{7}$$

On the r. h. s. of eq. (6), the sum of the first and third terms corresponds to the x-component of the flows $\{ \underline{\Gamma}^a_\perp \}_{coll}$ in eq. (1) and the sum of others to that of $\{ \underline{\Gamma}^a_\perp \}_{fluct}$. Lately, Wimmel [21] has proposed that the third term, related to the anisotropy of pressure tensor, becomes important in the cross-magnetic diffusion in a fully ionized plasma. In the weakly ionized plasma, however, this third term is not important as pointed out by its ratio to the first term, $< \alpha > S_e \ll 1$, where $< \alpha > = o\,(10^{-5} \sim 10^{-6})$ is given by Wimmel's estimation as the averaged ratio of the off-diagonal to diagonal terms in the pressure tensor and $S_e = o\,(10^2 \sim 10^3)$ for electrons in our plasma. The second term is about S_a times greater than the fourth one. In the turbulent state, this second term becomes dominant.

The field fluctuations $\delta \underline{E}\,(\underline{r}, t)$ are expressed by the density fluctuations $\delta n_a\,(\underline{r}, t)$ through Poisson's equation. On this basis of electrostatic field fluctuations, the cross moments $\overline{\delta n_a \, \delta E_\alpha}$ on the r. h. s. of eq. (6) are rewritten as

$$\overline{\delta n_a\,(\underline{r}, t)\; \delta E_\alpha\,(\underline{r}, t)} = \int d^3\underline{r}'\left(-\frac{\partial}{\partial \underline{r}_\alpha}\frac{1}{|\underline{r}' - \underline{r}|}\right) \sum_b \overline{\delta n_a\,(\underline{r}, t)\;\; \delta n_b\,(\underline{r}', t)} \tag{8}$$

The b-summation is carried out over all types of the charged particles. As known from eq. (8), the simultaneous secod moments $\overline{\delta n_a \; \delta n_b}$ in steady state are basic for determining the cross-magnetic diffusion flows due to the field fluctuations.

Hierarchy for the Moments of density Fluctuations

We derive an equation for the density fluctuations $\delta n_a\,(\underline{r}, t)$ from the continuity and momentum equations which are constructed from eqs. (3) and (5).

$$\frac{\partial}{\partial t}\,\delta n_a\,(\underline{r}, t) = L^a(\underline{r}) \cdot \delta n_a\,(\underline{r}, t) + G^a(\underline{r}/\underline{r}'') \cdot \sum_c z_c\;\; \delta n_e\,(\underline{r}'', t) +$$
$$- \sum_\alpha K^a_\alpha(\underline{r}) \cdot \left\{ \delta n_a\,(\underline{r}, t)\; \delta E_\alpha\,(\underline{r}, t) - \overline{\delta n_a\,(\underline{r}, t)\; \delta E_\alpha\,(\underline{r}, t)} \right\} \tag{9}$$

Here, $z_c = e_c/e$ is the valence of the c-type particles. The α-summation is carried out over the Cartesian coordinates x, y, z. The operators L^a, G^a and K^a_α at point $\underline{r}$ are defined by

$$L^a\,(\underline{r}) = \mu^a_\perp \left\{ \frac{Ta}{e_a}\left(\Delta_r + S_a^2 \frac{\partial^2}{\partial z^2}\right) - (1 + S_a^2)\, E_z \frac{\partial}{\partial z} \right\}$$

$$G^a(\underline{r}/\underline{r}'') = n_o(\underline{r})\; \mu^a_\perp \left\{ \left(\Delta_r + S_a^2 \frac{\partial^2}{\partial z^2}\right) + \varkappa \left(\frac{\partial}{\partial z} + S_a \frac{\partial}{\partial y}\right) \right\} \int \frac{d^3\underline{r}''}{|\underline{r}'' - \underline{r}|} \tag{10}$$

$$K^a_\alpha\,(\underline{r}) = \mu^a_\perp \left\{ \frac{\partial}{\partial r_\alpha} + S_a \left(\frac{\partial}{\partial x}\,\delta y_\alpha - \frac{\partial}{\partial y}\,\delta x_\alpha\right) + S_a^2 \frac{\partial^2}{\partial z^2}\,\delta z_\alpha \right\}$$

where Δ_r is the Laplacian at the point $\underline{r}$, $\mu^a_\perp = (e_a/m_a V_a)/(1 + S_a^2)$ the transverse mobility, $\delta_{\alpha\,\beta}$ the Kronecker delta symbol, $\varkappa = n_o^{-1}\,\partial n_o/$ $\varkappa$ the reciprocal of the characteristic dimension of the plasma and $n_o(\underline{r}) = \int d^3\underline{v}\, f_a\,(\underline{r}, \underline{v})$

the charged particle density. Here, we assumed $n_a(\underline{r}) = n_b(\underline{r})$ and abbreviated them by $n_o(\underline{r})$. In deriving the momentum equations, we have made the following two assumptions

a) $\int d^3\underline{v}\, v_\alpha\, v_\beta\, \delta Na(\underline{r}, \underline{v}, t) \simeq \delta_{\alpha\beta} \frac{Ta}{m_a} \delta na(\underline{r}, t)$

where Ta is the temperature of the charged particles and

b) we neglect the time derivative of $\int d^3\underline{v}\, \underline{v}\, \delta Na(\underline{r}, \underline{v}, t)$.

Assumption a) follows from the discussions at eq. (6) and assumption b) is valid for the high collision frequencies ν_a.

From eq. (9) we construct an equation for the second moments $\overline{\delta n_a\, \delta n_b}$ in steady state

$$0 = \left\{ L^a(\underline{r}) + L^b(\underline{r}')\right\} \cdot \overline{\delta n_a(\underline{r})\, \delta n_b(\underline{r}')} + G^a(\underline{r}/\underline{r}'') \cdot \sum_c z_c\, \overline{\delta n_c(\underline{r}'')\, \delta n_b(\underline{r}')} +$$
$$+ G^b(\underline{r}'/\underline{r}'') \cdot \sum_c z_c\, \overline{\delta n_a(\underline{r})\, \delta n_c(\underline{r}'')} - \sum_\alpha \left\{K^a_\alpha(\underline{r}) \cdot {}^a_\alpha V^b(\underline{r}, \underline{r}') + \right.$$
$$\left. + K^b_\alpha(\underline{r}') \cdot {}^aV^b_\alpha(\underline{r}, \underline{r}')\right\}, \tag{11}$$

which contains the simultaneous third moments defined by

$${}^a_\alpha V^b(\underline{r}, \underline{r}') = \overline{\delta n_a(\underline{r}, t)\ \delta E_\alpha(\underline{r}, t)\ \delta n_b(\underline{r}', t)} \tag{12a}$$

$${}^aV^b_\alpha(\underline{r}, \underline{r}') = \overline{\delta n_a(\underline{r}, t)\ \delta E_\alpha(\underline{r}', t)\ \delta n_b(\underline{r}', t)}. \tag{12b}$$

Equation (11) is not closed, because of the appearance of third moments. We truncate this hierarchy by some appropriate approximations. At the next section we first treat the case of small amplitude fluctuations, in which the third moments (12) and higher order ones are completely negligible.

The second moments $\overline{\delta n_a\, \delta n_b}$ are connected with the spatial correlation function χ_{ab}.

$$\overline{\delta n_a(\underline{r})\ \delta n_b(\underline{r}')} = \chi_{ab}(\underline{r}, \underline{r}') + \delta_{ab}\ \delta(\underline{r} - \underline{r}')\ n_a(\underline{r}), \tag{13}$$

where $\delta(\underline{r}-\underline{r}')$ is the Dirac delta function. Hereafter we omit the argument t in the correlation functions (cf. eqs. (11) - (13)).

Effect of small amplitude Fluctuations

The fluctuations are assumed to be homogeneous. This assumption is valid only if the mean density gradient is weak enough. The diffusion coefficient due to the small amplitude fluctuations is expressed in a form proportional to B^{-1}. However, the numerical factor A defined in eq. (2) is of the order $(1/n_o h_e^3)$, which means inversely proportional to the number of charged particles in the electron Debye sphere. These small values of A indicate that the diffusion coefficients due to the fluctuations are small compared to those due to the Boltzmann type collisions even for strong magnetic fields

B with $S_e = o(10^2 \sim 10^3)$. For reason of studying the influence of the correlation functions it is usefull to begin with that easiest case.

From the assumption of homogeneous fluctuations, the second moments $\overline{\delta N_a \delta N_b}$ depend on the coordinate difference $\underline{R} = \underline{r}' - \underline{r}$ only. Then, the correlation functions are expressed as

$$\chi_{ab}(\underline{r}, \underline{r}') = \chi'_{ab}(\underline{R}) \,. \tag{14}$$

Applying the Fourier-transformation

$$\tilde{\chi}_{ab}(\underline{k}) = \int d^3\underline{R}\; e^{i\underline{k}\cdot\underline{R}}\, \chi'_{ab}(\underline{R}) \tag{15}$$

to eq. (11) and taking account of eq. (13), we obtain simultaneous algebraic equations for the Fourier-transforms $\tilde{\chi}_{ab}(\underline{k})$.

$$0 = \left\{ \tilde{L}^a_o(\underline{k}, \underline{r}) + \tilde{L}^b_o(\underline{k}, \underline{r}') \right\} \tilde{\chi}_{ab}(\underline{k}) + \tilde{G}^a_o(\underline{k}, \underline{r}) \sum_c z_c\, \tilde{\chi}_{cb}(\underline{k}) + \tilde{G}^b_o(\underline{k}, \underline{r}') \sum_c z_c\, \tilde{\chi}_{ac}(\underline{k}) + \\ + \left\{ z_b\, \tilde{G}^a_o(\underline{k}, \underline{r}) + z_b\, \tilde{G}^b_o(\underline{k}, \underline{r}') \right\} n_o(\underline{r}) \tag{16}$$

The coefficients $\tilde{L}^a_o$, $\tilde{G}^a_o$ are the Fourier-transforms of the operators in eq. (10). We can readily find the identities.

$$\tilde{L}^{a\,*}_o(\underline{k}, \underline{r}') = \tilde{L}^a_o(\underline{k}, \underline{r}) \,, \qquad \tilde{G}^{a\,*}_o(\underline{k}, \underline{r}') = \tilde{G}^a_o(\underline{k}, \underline{r}) \,, \tag{17}$$

where the star * denotes the complex conjugate.

Comparing eq. (16) with the corresponding equation given by substituting the suffix $a \rightleftarrows b$ and by taking the complex conjugate, we obtain a simple relation

$$\tilde{\chi}_{ab}(\underline{k}) = \tilde{\chi}^*_{ba}(\underline{k}) \,. \tag{18}$$

This relation is usefull to solve the simultaneous algebraic equations (16). The reason for the Hermitian symmetry of eq. (18) is that the inertia effect of charged particles in their density fluctuations was neglected in deriving eq. (9). We further obtain another relation, (21), for $\tilde{\chi}_{ab}$ from the following equation (19) by taking account of the fact that the cross moments $\overline{\delta n_a \delta E_\alpha}$ are real. By using the Fourier-transforms of χ_{ab}, the cross moments (8) are expressed by

$$\overline{\delta n_a(r)\, \delta E_\alpha(r)} = i \frac{e}{2\pi^2} \int d^3\underline{k}\, \frac{k_\alpha}{k^2} \sum_b Z_b\, \tilde{\chi}_{ab}(\underline{k}) \tag{19}$$

$$= - \frac{e}{2\pi^2} \int d^3\underline{k}\, \frac{k_\alpha}{k^2} \,\mathrm{Im}\left\{ \tilde{\chi}_{ei}(\underline{k}) \right\} \,, \tag{20}$$

where Im{" } repressents the imaginary part. In deriving eq. (20) from eq. (19), we used the relation (18) and

$$\tilde{\chi}_{ai} + \tilde{\chi}^*_{ai} = \tilde{\chi}_{ae} + \tilde{\chi}^*_{ae} \tag{21}$$

which was mentioned before. As known from eq. (20), the cross moments are independent of the type of charged particles. This leads to the fact that the flows $\{\Gamma_1^a\}_{fluct}$ in strong magnetic fields $\underline{B}$ are of the same order for ions and for electrons (See eq. (6)).

Using relations (18) and (21), we solve eq. (15). The imaginary parts of $\tilde{\chi}_{ei}$ in the integrand of eq. (20) are given by

$$\mathrm{Im}\{\tilde{\chi}_{ei}(\underline{k})\} = \frac{\mathrm{Im}\{(\tilde{L}_o^i + \tilde{L}_o^{e*})(\tilde{G}_o^{i*} - \tilde{G}_o^e)\}\, n_o(\underline{r})}{|\tilde{L}_o^i + \tilde{L}_o^{e*}|^2 + \mathrm{Re}\{(\tilde{L}_o^i + \tilde{L}_o^{e*})(\tilde{G}_o^{i*} - \tilde{G}_o^e)\}} . \tag{22}$$

In order to calculate the r. h. s., we make the following approximations

$$(\varkappa h_e)\left(\frac{h_e e E_z}{Te}\right) \ll 1 \quad \text{and} \quad (k_z h_e)\left(\frac{h_e e E_z}{Te}\right) \ll 1 . \tag{23}$$

These are based on the assumptions $(h_e e E_z/Te) \ll 1$. In the plasma considered here, the electron Debye length h_e is comparable to the mean free path of charged particles, because of $\nu_a \simeq \pi_a$. The energy gain of an electron over one mean free path is assumed to be small compared to its thermal energy. After some straightforward but lengthy calculations, we obtain values of the r. h. s. of eq. (22) which are proportional to $\varkappa$, i. e., to the mean density gradient. These values lead to

$$A = \frac{\pi}{4} \frac{r_o}{h_e} , \tag{24}$$

where $r_o = e^2/Te$ is the classical electron radius. We have ignored the fourth term in eq. (6), because it is about S_a times less than the second term in strong magnetic fields $\underline{B}$. The magnitudes of A are of the order $(1/n_o h_e^3)$. Therefore, the diffusion coefficients are small, maximally of the same order as those due to the Boltzmann type collisions.

If the mean density of charged particles is uniform, i. e., $\varkappa = 0$, the values of the imaginary parts of $\tilde{\chi}_{ab}$ become zero. Naturally it follows that the cross magnetic flows vanish. In this case, the functions $\tilde{\chi}_{ei}$ are real and given by

$$\tilde{\chi}_{ei}(\underline{k}) = \frac{n_o}{k^2 h_e^2} , \quad \text{for } Ti = Te. \tag{25}$$

The above results for $\tilde{\chi}_{ei}$ differ from the thermal equilibrium calculation, $n_o/(1 + k^2 h_e^2)$. The expression (25) has no shielding effect by other charged particles. This shielding effect is removed by many collisions by the neutrals.

Current convective Instability

Before discussing the turbulent state, we examine the current convective instability, which was first studied by Kadomtsev.[2] His theory is based on the approximation of quasi-neutrality $\delta n_i \simeq \delta n_e$. Now we take account of the effect of the ion temperature. We use the linearized eq. (9), in which the quasi-neutrality condition is not necessary. Applying the Fourier-transformation to that equation, we obtain

$$-i\omega\,\delta\tilde{n}_a(\underline{k},\omega) = \tilde{L}^a_o(\underline{k},\underline{r})\,\delta\tilde{n}_a(\underline{k},\omega) + \tilde{G}^a_o(\underline{k},\underline{r})\sum_b z_b\,\delta\tilde{n}_b(\underline{k},\underline{r}) \qquad (26)$$

The coefficients $\tilde{L}^a_o$, $\tilde{G}^a_o$ are the same as in eq. (16) and

$$\delta\tilde{n}_a(\underline{k},\omega) = \int dt \int d^3\underline{r}\; e^{i\omega t - i\underline{k}\cdot\underline{r}}\,\delta n_a(\underline{r},t) \qquad (27)$$

being the definition of Fourier-transformation.

In the simultaneous equations (26) for $\delta\tilde{n}_i$ and $\delta\tilde{n}_e$, a determinant of the coefficients is given by

$$E(\underline{k},\omega;\underline{r}) = (\tilde{L}^i_o + i\omega + \tilde{G}^i_o)(\tilde{L}^e_o + i\omega - \tilde{G}^e_o) + \tilde{G}^i_o\,\tilde{G}^e_o\,, \qquad (28)$$

which corresponds to the dielectric constant. Setting this ε equal to zero, we obtain the dispersion relation for the current convective instability waves.

$$0 = \left\{ \frac{1}{1+S_i^2}\frac{k_\perp^2}{k^2} + \left|\frac{\mu i}{\mu e}\right|\frac{k_z^2}{k^2} + i\frac{1}{1+S_i^2}\frac{\varkappa}{k^2}\left(\frac{1}{S_i}\frac{k_y}{k} - \frac{k_z}{k}\right)\right\}\omega +$$

$$- (u_z^e k_z - i D_e k_z^2)\left\{\left(1+\frac{Ti}{Te}\right)\left(\frac{1}{1+S_i^2}\frac{k_\perp^2}{k^2} + \frac{k_z^2}{k^2}\right) - i\frac{1}{1+S_i^2}\frac{\varkappa}{k}\frac{k_x+S_i k_y}{k}\right\} \qquad (29)$$

Here, $k_\perp^2 = k_x^2 + k_y^2$ the square of the component of $\underline{k}$ perpendicular to $\underline{B}$ and $u_z^e = \mu_e E_z$ the electron velocity due to the external electric field E_z along the z-direction. We have assumed that the wave lengths are large compared to the electron Debye length h_e. The relation (29) well agrees with eq. (4) of Kadomtves's paper[2] at the limit $Ti/Te \to 0$. Under the condition $(kz/k)^2 \ll 1$ in eq. (29), the real parts of the complex frequencies $\omega = \omega_o + i\gamma$ are given by

$$\omega_o = \frac{1}{a}\,u_z^e\,k_z\,, \qquad (30)$$

where $a = 1 + (1+S_i^2)\,|\mu e|\,k_z^2/(\mu_i\,k_\perp^2)$. The above values of ω_o are considered to be in the frequency range $(10^4 - 10^6)\,H_z$, and correspond to those of the helical wave in the positive column. The imaginary parts are given by

$$\gamma = -\frac{1}{a}\left\{\frac{k_x + S_i k_y}{k_\perp} + \frac{1}{a}\left(\frac{1}{S_i}\frac{k_y}{k_\perp} - \frac{k_x}{k_\perp}\right)\right\}\frac{\varkappa}{k_\perp}u_z^e k_z +$$

$$-\frac{1}{a}\left(1+\frac{Ti}{Te}\right)\left\{1 + (1+S_i^2)\frac{k_z^2}{k_\perp^2}\right\}D_e k_z^2\,, \qquad (31)$$

where $D_e = Te/(m_e \nu_e)$ is the electron diffusion coefficient due to the Boltzmann type collisions. The first term on the r. h. s. of eq. (31) represents the growth rate of the current convective instability and the second term represents the damping by the electron diffusion along the magnetic field $\underline{B}$.

As known from eqs. (30) and (31), the ion temperature has an additional effect on the diffusion damping. The maximum growth rate is given for the wave vectors $\underline{k}$ nearly in the direction

$$\tan \varphi = S_i \tag{32}$$

where φ is the angle between the perpendicular component $k_\perp$ and the x-axis. The wave fields $\delta \underline{E}$ whose wave vectors $\underline{k}$ are subject to the condition (32) perturb the ion velocities mainly in the x-direction. These velocity perturbations are most effective to peturb the density of charged particles, because its mean density has a gradient in the x-direction.

Effect of large amplitude Fluctuations

When the amplitudes of instability waves grow, many other modes are induced by non-linear effects. The small scale modes die out by the large damping effect which results from the diffusion along the magnetic field $\underline{B}$ (See the second term of eq. (31)). We consider the turbulent state in which the field fluctuations become stationary and homogeneous. This state corresponds to that of the positive column in strong magnetic fields with B several times greater than the critical field B_c. In this turbulent region, the discharge electric field remains constant for increasing magnetic fields. The diffusion flows are dominantly due to the field fluctuations. Therefore, we can not neglect the third order moments in eq. (11). The third order moments are determined by fourth order moments, the fourth order moments by next higher order ones, and so on. For a calculation, we must truncate this hierarchy. Then, we apply a method similar to Milljonshchikov's to the fourth order moments, mentioned before. The validity of this method will be discussed later.

Applying the Fourier-transformation to eq. (11), we obtain

$$0 = \left\{ \tilde{L}_o^a + \tilde{L}_o^{b*} \right\} \tilde{\chi}_{ab}(\underline{k}) + \tilde{G}_o^a(\underline{k}) \sum_c \tilde{\chi}_{cb}(\underline{k}) + \tilde{G}_o^{b*}(\underline{k}) \sum_c z_c \tilde{\chi}_{ac}(\underline{k}) +$$

$$- \sum_\alpha \left\{ \tilde{K}_\alpha^a(\underline{k})\, {}^a_\alpha\tilde{V}^b(\underline{k}) + \tilde{K}_\alpha^{b*}(\underline{k})\, {}^a\tilde{V}_\alpha^{\,b}(\underline{k}) \right\} \tag{33}$$

The third order moments are expressed by functions of the coordinate difference $\underline{R}$ only, assuming homogeneous turbulence. We have neglected the second term on the r. h. s. of eq. (13) which was important in for the small amplitude fluctuations.

In order to obtain $\tilde{\chi}_{ab}$, we have to determine the Fourier-transforms of the third order moments ${}^a_\alpha V^b(\underline{K})$ and ${}^aV^b_\alpha(\underline{R})$. We first construct equations for the ${}^a_\alpha V^b$ by multiplying eq. (9) for the b-type particles at the point $\underline{r}'$by $\delta n_a(\underline{r}, t)\, \delta E_\alpha(\underline{r}, t)$ and then by averaging them.

$$i\omega\, {}^a_\alpha V^b(\underline{r}) = L_o^b(\underline{r}') \cdot {}^a_\alpha V^b(\underline{R}) + G_o^b(\underline{r}'|\underline{r}'') \cdot \sum_c z_c\, {}^a_\alpha V^b(\underline{R}'') -$$

$$- \sum_\beta K_\beta^b(\underline{r}') \cdot {}^a_\alpha W_\beta^b(\underline{r}, \underline{r}'), \tag{34}$$

where $\underline{R}'' = \underline{r}'' - \underline{r}'$ and $i\omega$ stands for the time differential at the point $\underline{r}'$. The simultaneous fourth order moments are defined as

$${}^a_\alpha W_\beta^b(\underline{r}, \underline{r}') = \overline{\delta n_a(\underline{r})\, \delta E_\alpha(\underline{r}) \left\{ \delta n_b(\underline{r}')\, \delta E_\beta(\underline{r}') - \overline{\delta n_b(\underline{r}')\, \delta E_\beta(\underline{r}')} \right\}} \tag{35}$$

By taking a = e for electrons and b = i or b = e for ions or electrons in eq. (34), respectively, we get after Fourier-Transformation one set of equations for ${}^{e}_{\alpha}V^{i}$ and ${}^{e}_{\alpha}V^{e}$. One formal solution of them is given by

$$ {}^{e}_{\alpha}\tilde{V}^{i}(\underline{k}) = \left\{ (\tilde{L}^{e\,*}_{o} - i\omega - \tilde{G}^{e\,*}_{o})\, \tilde{K}^{i\,*}_{\beta}\, {}^{e}_{\alpha}\tilde{W}^{i}_{\beta} + \tilde{G}^{i\,*}_{o}\, \tilde{K}^{e}_{\beta}\, {}^{e}_{\alpha}\tilde{W}^{e}_{\beta} \right\} / \varepsilon^{*}(\underline{k},\omega) \,. \tag{36a} $$

The repeated Greek letter indicates summation over the x, y, z components. From another set of equations for ${}^{i}_{\alpha}V^{i}$ and ${}^{i}_{\alpha}V^{e}$, which is given by putting a = i and b = i or b = e in eq. (34), we obtain

$$ {}^{i}_{\alpha}\tilde{V}^{e}(\underline{k}) = \left\{ (\tilde{L}^{i\,*}_{o} - i\omega + \tilde{G}^{i\,*}_{o})\, \tilde{K}^{e\,*}_{\beta}\, {}^{i}_{\alpha}\tilde{W}^{e}_{\beta} - \tilde{G}^{e\,*}_{o}\, \tilde{K}^{i\,*}_{\beta}\, {}^{i}_{\alpha}\tilde{W}^{i}_{\beta} \right\} / \varepsilon^{*}(\underline{k},\omega) \tag{36b} $$

Secondly, we construct the corresponding equations for the other third order moments ${}^{a}V^{b}_{\alpha}$ by multiplying eq. (9) for the a-type particles at the point $\underline{r}$ by $\delta n_b(\underline{r}')\ \delta E_{\alpha}(\underline{r}')$ and by averaging them.

$$ -i\omega\ {}^{a}V^{b}_{\alpha}(\underline{R}) = L^{a}_{o}(\underline{r}) \cdot {}^{a}V^{b}_{\alpha}(\underline{R}) + \tilde{G}^{a}_{o}(\underline{r}|\underline{r}'') \cdot \ z_c\ {}^{c}V^{b}_{\alpha}(\underline{R}'') - K^{a}_{\beta}(\underline{r}) \cdot {}^{a}_{\alpha}W^{b}_{\beta}(\underline{r},\underline{r}') . \tag{37} $$

Similarly to eq. (36), we obtain the following two formal solutions.

$$ {}^{e}\tilde{V}^{i}_{\alpha}(\underline{k}) = \ (\tilde{L}^{i}_{o} + i\omega + \tilde{G}^{i}_{o})\, \tilde{K}^{e}_{\beta}\, {}^{e}_{\beta}\tilde{W}^{i}_{\alpha} - \tilde{G}^{e}_{o}\, \tilde{K}^{i}_{\beta}\, {}^{i}_{\beta}\tilde{W}^{i}_{\alpha} \ / \ \varepsilon\,(\underline{k},\omega) \tag{38a} $$

$$ {}^{i}\tilde{V}^{e}_{\alpha}(\underline{k}) = \ (\tilde{L}^{e}_{o} + i\omega - \tilde{G}^{e}_{o})\, \tilde{K}^{i}_{\beta}\, {}^{i}_{\beta}\tilde{W}^{e}_{\alpha} + \tilde{G}^{i}_{o}\, \tilde{K}^{e}_{\beta}\, {}^{e}_{\beta}\tilde{W}^{e}_{\alpha} \ / \ \varepsilon\,(\underline{k},\omega) \tag{38b} $$

Comparing the above two type solutions (36) and (38), we find readily the following relation

$$ {}^{a}_{\alpha}\tilde{V}^{b}(\underline{k}) = {}^{b}\tilde{V}^{a\,*}_{\alpha}(\underline{k}), \tag{39} $$

if ${}^{a}_{\alpha}\tilde{W}^{b}_{\beta} = {}^{b}_{\beta}\tilde{W}^{a\,*}_{\alpha}$, which follows from the symmetry in homogeneous turbulence. Hence, we obtain the imaginary parts of $\tilde{\chi}_{ei}$ from eq. (33) by using the relations (18), (21) and (39).

$$ \mathrm{Im}\left\{ \tilde{\chi}_{ei}(\underline{k}) \right\} = \frac{\mathrm{Im}\left\{ (\tilde{L}^{i}_{o} + \tilde{L}^{e\,*}_{o})\,(\tilde{K}^{e}_{\alpha}\, {}^{e}_{\alpha}\tilde{V}^{i} + \tilde{K}^{i\,*}_{\alpha}\, {}^{e}\tilde{V}^{i}_{\alpha}) \right\}}{\left|\tilde{L}^{i}_{o} + \tilde{L}^{e\,*}_{o}\right|^{2} + \mathrm{Re}\left\{ (\tilde{L}^{i}_{o} + \tilde{L}^{e\,*}_{o})\,(\tilde{G}^{i\,*}_{o} - \tilde{G}^{e}_{o}) \right\}} \tag{40} $$

This is quite parallel to the solution (22) for the small amplitude fluctuations. They are both derived from eq. (11). The solution (22) is based on the neglection of third order moments, whereas the solution (40) is based on the neglection of the second term on the r. h. s. of eq. (13). These two different approximations lead to the different factors $(\tilde{G}^{i\,*}_{o} - \tilde{G}^{e}_{o})\, n_o$ and $(\tilde{K}^{e}_{\alpha}\, {}^{e}_{\alpha}\tilde{V}^{i} + \tilde{K}^{i\,*}_{\alpha}\, {}^{e}\tilde{V}^{i}_{\alpha})$ in the numerators of solutions (22) and (40).

In order to calculate the r. h. s. of eq. (40), we assume that $\langle k_x^2 \rangle \simeq \langle k_y^2 \rangle \simeq \langle k_z^2 \rangle$ and $\langle k^2 \rangle^{1/2} = o\,(h_e)$, where < "> means averaging. Under these assumptions, the quantity $\varepsilon\,(\underline{k},\omega)$ of eq. (28) is approximately expressed by the dimensionless form

$$ \frac{\varepsilon\,(\underline{k},\omega)}{(4\pi\, n_o\, e\, \mu_e)^2} \simeq \left|\frac{\mu_i}{\mu_e}\right| \left(\frac{1}{1+S_i^2}\frac{k_\perp^2}{k^2} + \frac{k_z^2}{k^2} \right) \frac{k_z^2}{k^2}\ x^2 \tag{41} $$

where $x = k h_e$. The r. h. s. corresponds to the diffusion damping terms in eq. (29). By using eqs. (36) and (38), the numerator on the r. h. s. of eq. (40) is written as

$$\mathrm{Im}\left\{ (\tilde{L}_o^i + \tilde{L}_o^{e*}) (\tilde{K}_\alpha^e \, {}_\alpha^e\tilde{V}^i + \tilde{K}_\alpha^{i*} \, {}^e\tilde{V}_\alpha^i) \right\} =$$

$$= \mathrm{Im}\left\{ (\tilde{L}_o^i + \tilde{L}_o^{e*}) (\tilde{K}_\alpha^e \tilde{K}_\beta^{i*} \left[\frac{\tilde{L}_o^{e*} - i\omega - \tilde{G}_o^{e*}}{\varepsilon *} + \frac{\tilde{L}_o^i + i\omega + \tilde{G}_o^i}{\varepsilon} \right] {}_\alpha^e\tilde{W}_\beta^i + \right.$$

$$\left. + \frac{\tilde{G}_o^{i*} \tilde{K}_\alpha^e \tilde{K}_\beta^e}{\varepsilon *} {}_\alpha^e\tilde{W}_\beta^e - \frac{\tilde{G}_o^e \tilde{K}_\alpha^{i*} \tilde{K}_\beta^i}{\varepsilon} {}_\beta^i\tilde{W}_\alpha^i \;) \right\}$$

$$\simeq (4\pi n_o e \mu_e)^2 \frac{\varkappa}{k} \left(\frac{k_z}{4\pi n_o e}\right)^2 \frac{\mathrm{Si}\frac{k_y}{k}}{1+S_i^2\frac{k_z^2}{k^2}} \mathrm{Re}\left\{ {}_z^e\tilde{W}_z^e (\underline{k}) \right\}. \tag{42}$$

In deriving the r. h. s. of the above equation, we used the same approximations as in eq. (41) and omitted the terms which vanish by the $\underline{k}$-integration of eq. (20). The reason why the z-components of ${}_\alpha^a\tilde{W}_\beta^b (\underline{k})$ are kept is: that the z-components of $\tilde{K}_\alpha^a (\underline{k})$ are main terms in the strong magnetic fields $\underline{B}$. We have neglected the imaginary part of ${}_z^e\tilde{W}_z^e$, because of the assumption that the turbulent fluctuations are nearly isotropic. The denominator on the r. h. s. of eq. (40) is given by

$$\left| \tilde{L}_o^i + \tilde{L}_o^{e*} \right|^2 + \mathrm{Re}\left\{ (\tilde{L}_o^i + \tilde{L}_o^{e*}) (\tilde{G}_o^{i*} - \tilde{G}_o^e) \right\} \simeq (4\pi n_o e \mu e)^2 \left(\frac{k_z}{k}\right)^2 (1+x^2) x^2. \tag{43}$$

Hence, the equation (40) becomes

$$\mathrm{Im}\left\{ \tilde{\chi}_{ei}(k) \right\} = \frac{\varkappa k_y}{(4\pi n_o e)^2} \left(\frac{k}{k_z}\right)^2 \frac{1}{(1+x^2) x^2} \frac{\mathrm{Si}}{1+S_i^2\frac{k_z^2}{k^2}} \mathrm{Re}\left\{ {}_z^e\tilde{W}_z^e \right\}. \tag{44}$$

It is worthwhile to note that the r. h. s. is proportional to $\varkappa$, i. e., to the mean density gradient. The above imaginary parts determine the amounts of diffusion flows $\{\Gamma_x{}^a\}_{fluct}$ through eqs. (6) and (20). The proportionality to the mean density gradient is explained as follows: the flow $\{\Gamma_x{}^a\}_{fluct}$ is given as a net ($\delta\underline{E} \times \underline{B}$) drift of charged particles. As the fluctuation fields $\delta\underline{E}$ are nearly isotropic, then the net flow depends on the nonuniformity of the density in the positive and the negative x-direction.

We express the fourth order moments ${}_z^eW_z^e (\underline{r}, \underline{r}')$, which are defined by eq. (35), approximately by the sum of the products of second moments according to Millionshchikov's method.[18]

$${}_z^eW_z^e (\underline{r}, \underline{r}') = \overline{\delta n_e(\underline{r})\, \delta n_e(\underline{r}')} \cdot \overline{\delta E_z(\underline{r})\, \delta E_z(\underline{r}')} +$$

$$+ \overline{\delta n_e(\underline{r})\, \delta E_z(\underline{r}')} \cdot \overline{\delta n_e(\underline{r}')\, \delta E_z(\underline{r})}\,. \tag{45}$$

Fortunately in the present case the term $\overline{\delta n_e(\underline{r})\,\delta E_z(\underline{r})} \cdot \overline{\delta n_e(\underline{r}')\,\delta E_z(\underline{r}')}$, which does not vanish at the limit $(\underline{r}' - \underline{r}) \to \infty$, was cancelled out in the expression (45). Millionshchikov's method is orgininally valid for normal distributions of fluctuations. However, this method is still true, even if the third order moments do not vanish.[18)~20)] The second term on the r. h. s. of eq. (45) is the product of the cross moments, and therefore, it is of the order $\varkappa^2$. By neglecting this second term, we express the Fourier-transforms of the fourth order moments as follows:

$$
{}^{e}_{z}\tilde{W}^{e}_{z}(\underline{k}) = \int d^3\underline{r}'\, e^{i\underline{k}\cdot(\underline{r}' - \underline{r})}\; {}^{e}_{z}W^{e}_{z}(\underline{r}, \underline{r}')
$$

$$
\simeq \int d^3\underline{r}'\, e^{i\underline{k}\cdot(\underline{r}' - \underline{r})}\; \overline{\delta n_e(\underline{r})\,\delta n_e(\underline{r}')} \cdot \overline{\delta E_z(\underline{r})\,\delta E_z(\underline{r}')}
$$

$$
= \overline{(\delta n_e)^2}\;\overline{(\delta E_z)^2} \int d^3\underline{R}\, e^{i\underline{k}\cdot\underline{R}}\; \Psi(\underline{R})\,. \qquad (46)
$$

Here, $\Psi(\underline{R})$ is a function denoting the spatial dependence of $\overline{\delta n_e(\underline{r})\,\delta n_e(\underline{r}')\,\delta E_z(\underline{r})\,\delta E_z(\underline{r}')}$. From the assumption of the nearly isotropic fluctuations, the Fourier-transforms (46) are treated as real.

The diffusion flows $\{\Gamma_x{}^a\}_{fluct}$ are mainly given by the second term in eq. (6), supplemented by eqs. (20), (44) and (46). After some calculations, we obtain the following result for the numerical factor A.

$$
A = \frac{\overline{(\delta n_e)^2}}{n_o^2}\;\frac{\overline{|\delta E|^2}}{N_o\, Te}\left\{\frac{1}{32\pi^2}\,\frac{1}{h_e^3}\int_0^\infty \frac{dx}{1+x^2}\int d^3\underline{R}\, e^{i\underline{k}\cdot\underline{R}}\;\Psi(\underline{R})\right\}. \qquad (47)
$$

The first factor on the r. h. s. is the ratio of the mean square amplitude of electron density fluctuations to that of the mean density, the second the ratio of the field fluctuation energy to the particle kinetic energy in unit volume, and the third is related to the correlation length of the turbulent fluctuations. If we take $\Psi(\underline{R}) = \exp\{-(R/\lambda_t)^2\}$, we obtain $(1/32)\,(\lambda_t/he)^2$ for the third factor. In the investigation of diffusion hitherto known, the values of A were represented by using the first or second, factor only.

Discussion

The turbulent diffusion coefficient is determined by three characteristic quantities given in eq. (47). For strong turbulence, it will be influenced not only by the density fluctuations, but also by the field fluctuations which result from the phase difference between ion and electron fluctuations. The turbulent correlation length λ_t is much smaller than the wave lengths of current convective instability waves, but larger than the electron Debye length h_e. The expression (47) is correct when $(\overline{|\delta E|^2}/n_o\, Te)$ is not so small compared to unity. In other words, the expressions (41) - (43) are good approximations when the values $\overline{(h_e\, e\, |\delta E|/Te)^2}$ are not so small as shown in eq. (23). We will here discuss the values of A for the intermediate region between the small and large amplitude fluctuations by comparing with the succesive method of Kadomtsev.[3)]

In this procedure, the following equations corresponding to eq. (26) are derived from eq. (9).

$$- i\omega \; \delta\tilde{n}_a(\underline{k}, \omega) = \tilde{L}^a_o(\underline{k}, \underline{r}) \; \delta\tilde{n}_a(\underline{k}, \omega) + \tilde{G}^{a'}_o(\underline{k}, \omega) \; \delta\tilde{V}(\underline{k}, \omega) + - \tilde{K}^a_\alpha(\underline{k}, \omega; \underline{r}) \sum_{\substack{k'+k'' = k \\ \omega'+\omega'' = \omega}} i k''_\alpha \Big\{ \delta\tilde{N}_a(\underline{k}', \omega') \delta\tilde{V}(\underline{k}'', \omega'') - \overline{\delta n_a(k', \omega') \delta V(k'', \omega'')} \Big\} \tag{48}$$

Here, $\delta\tilde{V}(\underline{k}, \omega)$ is the Fourier-transform of $\delta V(\underline{r}, t) = e \sum_b z_b \int d^3\underline{r}' \delta n_b(\underline{r}', t)/|\underline{r}' - \underline{r}|$ and the operator $\tilde{G}^{a'}_o$ does not contain the integral operation as in $\tilde{G}^a_o$ defined by eq. (10). We assume the quadratic terms on the r. h. s. of eq. (48) to be of next higher order. Making use of this assumption, we express $\delta\tilde{n}_a(\underline{k}, \omega)$ as a potential series in $\delta\tilde{V}(\underline{k}, \omega)$.

$$\delta\tilde{n}_a(\underline{k}, \omega) = G^a(\underline{k}, \omega) \; \delta\tilde{V}(\underline{k}, \omega) + - \tilde{K}^a_\alpha(\underline{k}, \omega; \underline{r}) \sum_{\substack{k'+k'' = k \\ \omega'+\omega'' = \omega}} i k''_\alpha \; G^a(\underline{k}'', \omega'') \Big\{ \delta\tilde{V}(\underline{k}', \omega') \; \delta\tilde{V}(\underline{k}'', \omega'') - \overline{\delta V(\underline{k}', \omega') \; \delta V(\underline{k}'', \omega'')} \Big\} + \text{third order terms of } \delta\tilde{V} + \ldots, \tag{49}$$

where $G^a(\underline{k}, \omega) = - \tilde{G}^{a'}_o(\underline{k}, \omega)/(i\omega + \tilde{L}^a_o)$.

The second-order moments of $\delta n_a(\underline{k}, \omega)$ are expressed by

$$\overline{\delta\tilde{n}_a(\underline{k}, \omega) \; \delta\tilde{n}_b(\underline{k}', \omega')} = \chi_{ab}(\underline{k}, \omega) \; \delta(\underline{k} + \underline{k}') \; \delta(\omega + \omega'), \tag{50}$$

which follows from the assumption of homogeneous fluctuations. The ω-integral of $\tilde{\chi}_{ab}(\underline{k}, \omega)$ gives the function $\tilde{\chi}_{ab}(\underline{k})$ in eq. (13) Multiplying $\delta\tilde{n}_a(\underline{k}, \omega)$ by $\delta\tilde{n}_b(\underline{k}', \omega')$ and using the relation (50), we obtain

$$\tilde{\chi}_{ab}(\underline{k}, \omega) = G^a(\underline{k}, \omega) \; G^{a*}(\underline{k}, \omega) \; (\delta\tilde{V})^2_{\underline{k}, \omega} + \ldots, \tag{51}$$

where $(\delta\tilde{V})^2_{k, \omega}$ is the Fourier-transform of the correlation function of field fluctuations. Similarly to eq. (44), the first term in eq. (51) leads

$$\operatorname{Im}\Big\{ \tilde{\chi}_{ei}(\underline{k}, \omega) \Big\} = \frac{\varkappa k_y}{k^2} \left(\frac{n_o e}{Te}\right)^2 \left\{ \frac{Si}{1+S_i^2 \frac{k_z^2}{k^2}} + \frac{Se}{1+S_e^2 \frac{k_z^2}{k^2}} \right\} (\delta\tilde{V})^2_{\underline{k}, \omega} \tag{52}$$

Integrating over ω the above equation (52) and substituting it into eq. (20), we obtain

$$A = \left(\frac{\lambda}{he}\right)^2 \left\{ \frac{e^2}{Te^2} \overline{(\delta V)^2} \right\}. \tag{53}$$

Here, $\overline{(\delta V)^2}$ is the mean square amplitude of field fluctuations and λ the correlation length defined by

$$\lambda = \left\{ \frac{1}{\overline{(\delta \tilde{V})^2}} \int d^3\underline{k} \frac{1}{k^2} \int d\omega \, (\delta \tilde{V})^2_{\underline{k},\omega} \right\}^{1/2} \tag{54}$$

The second factor on the r. h. s. of eq. (53), which is the relative fluctuation energy, corresponds to the second one in eq. (47). For the small amplitude fluctuations, this second factor is of the order $(1/n_o h_e^3)$ and λ is comparable to the electron Debye length h_e. Then, the values of A approximate those of eq. (24). For the strong turbulent fluctuations, however, the values of eq. (53) can not approach those of eq. (47). The reason is self-evident. When the higher order moments are necessary, the successive method is not tractable.

Acknowledgement

The author wishes to express his sincere thanks to Prof. Dr. G. Ecker and Prof. Dr. W. Kröll for their helpful discussions and many comments. He is indebted to Dr. K. H. Spatschek who gave him many discussions and suggestive comments. He is also grateful to Dr. H. Gratzl in the Kernforschungsanlage Jülich for valuable discussions. The author further gratefully acknowledges the support through the Landesamt für Forschung.

References

(1) L. Spitzer, Jr., Phys. Fluids 3, 695 (1960).

(2) B.B. Kadomtsev, Soviet Phys. Tech. Phys. 6, 927 (1962); Soviet Phys. JETP 16, 1191 (1963); J. Nucl. Energy C5, 31 (1963).

(3) B.B. Kadomtsev, "Plasma Turbulence", Academic Press (1965).

(4) A.A. Galeev, S.S. Moiseev and R.Z. Sagdeev, J. Nucl. Energy C6, 645 (1964).

(5) A.A. Galeev, Phys. Fluids 10, 1041 (1967).

(7) T.H. Dupree, Phys. Fluids 10, 1049 (1967); 9 1773 (1966)

(8) M. Matsumoto, J. Phys. Soc. Japan 26, 1280 (1969).

(9) F.C. Hoh, Rev. Mod. Phys. 34, 267 (1962).

(10) B.J. Eastlund, K. Josephy, R.F. Leheny and T.C. Marshall, Phys. Fluids 9, 2400 (1966).

(11) M. Porkoab and G.S. Kino, Phys. Fluids 11, 346 (1968)

(12) J. Walsh, S.P. Schlesinger, K. Josephy and T.C. Marshall, Phys. Fluids 12, 2374 (1969).

(13) E. Woehler, Phys. Fluids 10, 245 (1967).

(14) Yu.L. Klimontovich, "The Statistical Theory of Non-Equilibrium Processes in a Plasma", M.I.T. Press (1967)

(15) Kalman and Feix, "Nonlinear Effects in Plasma", London (1969) 239.

(16) Yu.L. Klimontovich and V. Ebeling, Soviet Phys. JETP 16, 104 (1963)

(17) L.D. Landau and E.M. Lifshitz, "Statistical Physics", Pergamon Press (1959) § 115.

(18) C.C. Lin, "Turbulent Flows and Heat Transfer", Princeton University Press (1959) 236.

(19) S. Chandrasekhar, Proc. Roy. Soc. A207, 301 (1951).

(20) G.K. Batchelor, "Homogeneous Turbulence", Cambridge University Press (1967).

(21) H.K. Wimmel, Nucl. Fusion 10, 117 (1970).

Forschungsberichte des Landes Nordrhein-Westfalen

Herausgegeben im Auftrage des Ministerpräsidenten Heinz Kühn
vom Minister für Wissenschaft und Forschung Johannes Rau

Sachgruppenverzeichnis

Acetylen · Schweißtechnik
Acetylene · Welding gracitice
Acétylène · Technique du soudage
Acetileno · Técnica de la soldadura
Ацетилен и техника сварки

Arbeitswissenschaft
Labor science
Science du travail
Trabajo científico
Вопросы трудового процесса

Bau · Steine · Erden
Constructure · Construction material · Soilresearch
Construction · Matériaux de construction · Recherche souterraine
La construcción · Materiales de construcción · Reconocimiento del suelo
Строительство и строительные материалы

Bergbau
Mining
Exploitation des mines
Minería
Горное дело

Biologie
Biology
Biologie
Biologia
Биология

Chemie
Chemistry
Chimie
Quimica
Химия

Druck · Farbe · Papier · Photographie
Printing · Color · Paper · Photography
Imprimerie · Couleur · Papier · Photographie
Artes gráficas · Color · Papel · Fotografía
Типография · Краски · Бумага · Фотография

Eisenverarbeitende Industrie
Metal working industry
Industrie du fer
Industria del hierro
Металлообработывающая промышленность

Elektrotechnik · Optik
Electrotechnology · Optics
Electrotechnique · Optique
Electrotécnica · Optica
Электротехника и оптика

Energiewirtschaft
Power economy
Energie
Energía
Энергетическое хозяйство

Fahrzeugbau · Gasmotoren
Vehicle construction · Engines
Construction de véhicules · Moteurs
Construcción de vehículos · Motores
Производство транспортных средств

Fertigung
Fabrication
Fabrication
Fabricación
Производство

Funktechnik · Astronomie
Radio engineering · Astronomy
Radiotechnique · Astronomie
Radiotécnica · Astronomía
Радиотехника и астрономия

Gaswirtschaft
Gas economy
Gaz
Gas
Газовое хозяйство

Holzbearbeitung
Wood working
Travail du bois
Trabajo de la madera
Деревообработка

Hüttenwesen · Werkstoffkunde
Metallurgy · Materials research
Métallurgie · Matériaux
Metalurgia · Materiales
Металлургия и материаловедение

Kunststoffe
Plastics
Plastiques
Plásticos
Пластмассы

Luftfahrt · Flugwissenschaft
Aeronautics · Aviation
Aéronautique · Aviation
Aeronáutica · Aviación
Авиация

Luftreinhaltung
Air-cleaning
Purification de l'air
Purificación del aire
Очищение воздуха

Maschinenbau
Machinery
Construction mécanique
Construcción de máquinas
Машиностроительство

Mathematik
Mathematics
Mathématiques
Matemáticas
Математика

Medizin · Pharmakologie
Medicine · Pharmacology
Médecine · Pharmacologie
Medicina · Farmacología
Медицина и фармакология

NE-Metalle
Non-ferrous metal
Metal non ferreux
Metal no ferroso
Цветные металлы

Physik
Physics
Physique
Física
Физика

Rationalisierung
Rationalizing
Rationalisation
Racionalización
Рационализация

Schall · Ultraschall
Sound · Ultrasonics
Son · Ultra-son
Sonido · Ultrasónico
Звук и ультразвук

Schiffahrt
Navigation
Navigation
Navegación
Судоходство

Textilforschung
Textile research
Textiles
Textil
Вопросы текстильной промышленности

Turbinen
Turbines
Turbines
Turbinas
Турбины

Verkehr
Traffic
Trafic
Tráfico
Транспорт

Wirtschaftswissenschaften
Political economy
Economie politique
Ciencias económicas
Экономические науки

Einzelverzeichnis der Sachgruppen bitte anfordern

Westdeutscher Verlag · Opladen

567 Opladen/Rhld., Ophovener Straße 1–3, Postfach 1620

MIX
Papier aus verantwortungsvollen Quellen
Paper from responsible sources
FSC® C105338

If you have any concerns about our products,
you can contact us on
ProductSafety@springernature.com

In case Publisher is established outside the EU,
the EU authorized representative is:
Springer Nature Customer Service Center GmbH
Europaplatz 3, 69115 Heidelberg, Germany

Printed by Libri Plureos GmbH
in Hamburg, Germany